CAPTURED TANKS UNDER THE GERMAN FLAG

RUSSIAN BATTLE TANKS

DR. WERNER REGENBERG

HORST SCHEIBERT

Front cover: see page 33.

Left page:
This T-34 (1943 model) was in service with the 6th Armored Division. Its running gear shows 1 each, fore and aft, later-type rubber-covered road wheel, and three road wheels between.

Schiffer Publishing Ltd

1469 Morstein Road, West Chester, Pennsylvania 19380

PHOTO CREDITS:

—Federal Archives (BA)
—Munin Publishers
—Barthols Collection
—Willy Brändlin
—Dr. Carl-Hermann Cutshorn
—Reinhard Frank
—Horst Riebenstahl
—Wolfgang Schneider
—Otto Strecker
—Rudi Weber
—Bernd Wittigayer

Translated from the German by Dr. Edward Force.

Copyright © 1990 by Schiffer Publishing.
Library of Congress Catalog Number: 89-063355.

Printed in the United States of America.
ISBN: 0-88740-201-1

A T-34 (1943 model) in service with Company 7 of Armored Regiment 11 in 1943. A German commander's cupola was mounted on the tank. In the middle of the bow is the regimental marking "Op" (Regimental Commander von Oppeln-Bronikowski). It is the same tank as was shown on the inside front cover.

CAPTURED TANKS

The term "captured tanks" refers to armored vehicles captured and reused by the capturing troops.

These tanks were either captured by the fighting forces in usable condition, or were quickly made usable, in order to replace their own losses or strengthen their fighting power. There were such tanks—quite a large number of them, in fact—used as individual units in very many groups, even within infantry divisions. The main thing was that drivers and crews could be found who could operate the weapons. But most of these tanks had a very short life, since spare parts and ammunition were lacking. In the latter case they were often used as towing tractors.

To avoid confusing them with enemy tanks, the captured tanks used at the front were usually painted with large-dimension German crosses or swastikas—on the bow as well.

Captured tanks also include those taken on the battlefield or in factories, which were overhauled and made usable by industries (including those in occupied countries) or larger military repair units. Entire combat units, from column to brigade strength, were equipped with these tanks.

The service of this type of troop unit was already a tradition in the German Army, since in World War I a lack of their own tanks (Tank Detachments 1 to 3 were equipped with the A7V tank) caused six tank units (Tank Detachments 11 to 16) to be set up with captured tanks.

In World War II these units were usually not used on the fighting front, but rather in rearward areas for guarding and training

purposes. Their own workshops as well as ammunition and spare-parts depots were maintained for them.

The Austrian and, in particular, the Czech tanks, which were taken over in great numbers after the union with Austria and the annexation of Czechoslovakian areas, are not to be regarded as captured tanks in the strict sense of the term.

Information on the weapons of other countries was gathered by the Army Weapons Office before and during the war, and so-called "foreign vehicle recognition manuals" were prepared. Here a system was used in which first the German name of the vehicle was given, followed by a number, and then letters in parentheses to indicate the land of origin. For battle tanks the numbers used were in the 700's, and so the official designation of the T-34, for example, was:

Battle Tank T-34 747 (r)

In this volume only captured Russian tanks will be covered which, after sometimes very minor work and the application of German elevation calibration, were put in service on the German side.

To take charge and make quick use of captured tanks, captured tank staffs were established to direct and oversee the whole handling of especially needed tanks. These staffs were ordered in July of 1941 to gather Russian T-35, T-28, BT and T-26 tanks. At that time the German side did not yet know about the KW-1 or even the T-34.

As before with Polish, Dutch, Belgian, English and particularly French tanks, the Army wanted to make the booty expected in the Russian campaign useful too.

But at the very beginning, major difficulties arose, since a great percentage of the tanks had been so badly damaged in battle that they could only be scrapped. A great many externally undamaged tanks were lacking necessary parts, which had either been destroyed by the Russians themselves or taken as replacement parts by the troops that captured the tanks.

Many of the tanks whose repair was at first regarded as worthwhile by the German Army, including the T-28 and T-35, weighed up to 52 tons and could scarcely be moved by German towing apparatus.

The main reason it was impossible to make large numbers of Russian tanks usable for the Wehrmacht was that German repair facilities were not even sufficient to repair damaged German tanks as quickly as necessary and get them back to the troops, or send them home for longer repairs. Thus by the end of October 1941 only about 100 Russian tanks could be put into service. Most of the Russian tanks lying around in the wide open spaces, estimated by the Wehrmacht at 10,000 units (!), remained where they were through the winter of 1941/42 and were usable only as scrap metal afterward.

Until the beginning of 1942, only a few BT and T-26 and even fewer T-60 tanks were repaired and made available to captured-tank units. Every one that was usable, especially the KW-1 and T-34, had already been taken along by the front-line troops and put into

Two T-26 tanks in a field workshop. In front is a T-26 B (1933 model) with Russian emblem and the inscription "captured 15th I.R.". Behind it is a T-26 E with a German cross, the name "Tiger II", and the tactical sign of the 3rd SS Division "Totenkopf". The T-26 E tanks were T-26 S types equipped with additional armor on the turret and body. (BA)

service. The T-34 and KW-1 were particularly prized pieces, since they strengthened the German armored defense against the Russian T-34 and KW-1 tanks.

Only as of mid-1942 did individual KW-1 and T-34 tanks come out of repair facilities on their way to German captured-tank units.

Most Russian tanks were overhauled in the tank repair facility in Riga and turned over to German units. As of 1943, individual T-34 tanks also came from the Daimler-Benz factory in Marienfelde and the Wumag firm in Görlitz.

The SS Armored Corps took over the tractor works (tank factory) in Kharkov after that city was retaken in the spring of 1943, and began the production of T-34 tanks there. The SS Armored Pursuit Detachment 2 of the Division "Das Reich" was equipped with these T-34's.

All in all, the number of Russian tanks reused out of the huge numbers captured was small. According to the officially conducted "Captured Tank Situation" study of May 1943, there were 63 Russian tanks (50 of them T-34), and in December 1944 53 Russian tanks (49 of them T-34) in service as battle or pursuit tanks in the Army and Waffen-SS.

But the report of the "Captured Tank Situation" was never complete, because various troop units, on account of their involvement in combat, never had a chance to make reports, and others kept an undeclared supply of tanks. More than 300 captured Russian battle tanks, though, were seldom in service on the German side at any one time.

In the winter of 1943/44 this T-34 was captured. The T-34/76 has been painted with a simple German cross and equipped with a German commander's cupola.

A light Russian T-26 (1933 model) in service with the 3rd SS Division "Totenkopf". The unit emblem, the death's-head, has been painted on the chain cover at the right front, and the SS runes are also added on the bow. The tank, with the strikingly big German cross, bears the name "Mistbiene" (manure bee). The T-26 (1933 model) was designated by the German Army as Armored Battle Vehicle T-26 B 738(r). (BA)

LIGHT RUSSIAN BATTLE TANKS

Among the vast number of tanks captured in the first months of the Russian campaign there were also a great many light battle tanks of the BT and T-26 types. The fact that by far the greatest numbers of these tanks were captured without any damage by enemy action suggested that the construction and preparation of these tanks must have been flawed. When used by German troops, the T-26 generally broke down because of clutch damage, hot steering apparatus or seized bearings, the BT because of too-weak shift linkages and faults in the electric system.

As of 1942 some T-60, and as of 1943 some T-70 tanks also saw service with German troop units as captured tanks. The T-60 and its successor, the T-70, were also used as tractors for Pak and artillery after their turrets were removed.

Upper right:
The same "Mistbiene" tank (see left page), but here abandoned by the German troops because of some defect. (Wittigayer)

Right:
Servicing a T-26 C somewhere on the eastern front. This tank has only a small cross painted on the upper body to mark it as a German vehicle. The T-26 was equipped with a 45-mm cannon. (Barthols)

Upper left:
In the winter of 1941/42 this T-26 C was photographed. The tank wears a simplified German cross on its turret and the lettering Fuka 10. (BA)

Above: The same tank. Its commander wears the armored uniform with fur cap, while some of the infantrymen wear snow coats. (BA)

Left:
Here a T-26 C tows a 3-ton Mercedes truck along a muddy path. The tank is marked with a cross and a swastika flag on the turret. The observation slit and pistol loophole in the turret are easy to see. (BA)

This T-26 S (1939 model) carries a big German cross on the turret and also a white number 3 on the left track cover.

These two photos show the
same Type T-26, also called
Armored Battle Vehicle T-26
C 740(r) by the German
Army, going into action
with infantry following it.
The lower picture shows that
a cross has also been painted
on the back of the turret. On
the tail the crew's gear,
ammunition boxes and fuel
canisters have been stowed.
(2 x BA)

BT-7

Right: This photo shows a BT-7 TU serving as a bunker to protect a fortified supply depot. This was the last task that captured tanks could fulfill if their motors or running gear could no longer be repaired.

Below: Apparently just painted with yellow camouflage paint, this BT-7, built in 1935, was photographed in 1943 or 1944. (BA)

A tank with a big German cross in a Russian city. At the right edge of the picture it can just be seen that the tank has a rear machine gun in its turret. This identifies it as a Type BT-7 M.

Three BT-7 tanks with very noticeable German crosses painted on. These were three different versions of the Type BT-7. In front in the picture, the last production type BT-7 M (also called BT-8) can be seen, with one oval and one round turret hatch as well as mountings for an anti-aircraft machine gun (1939 type). In the middle is the BT-7 (built in 1937), and at the rear a Type BT-7TU (built in 1935). The suffixed TU indicates that it is a command tank, thus it is equipped with a frame antenna. All three tanks have their original Russian shooting spotlights mounted. (Frank)

Left: This BT-742(r) is serving a final purpose. After seeing service on the German side, it serves now as a practice object for close-range tank fighting in an army school for paratroopers. The paratrooper is hanging two boxes of #24 explosive, connected by wire, on the barrel of the BT-7 M's gun. This operation, performed on an active tank, called for an ample portion of courage. (BA)

Captured T-60 tanks were seldom used as battle or reconnaissance vehicles in the German Army. Light tanks armed with machine guns or a 2-cm cannon were often rebuilt by the Wehrmacht as tractors or self-propelled guns. Thus most T-60 tanks, like this one with a built-on 2-cm cannon and a 1.IG in tow, were used as towing tractors. (BA)

13

T-70

Left:
The workshop company of the 82nd Infantry Division is repairing several T-70 and T-34 tanks to make up a captured tank unit. (BA)

Below:
Captured T-70 tanks found greater use as battle tanks, as they had an acceptable armament with their 47-mm cannon. But a great disadvantage was the one-man turret. This T-70 was in service with the 98th Infantry Division. (BA)

Above:
This T-70 is also in service with the 82nd Infantry Division. Here the tank's weapon is being tested or fired. (Strecker)

Upper right:
The T-70 as an ammunition tractor.

Right: Here a T-70 (without turret) is used as a towing tractor for a 7.5-cm Pak 40 antitank gun.

T-34/76
(Ausf. A-C)

MEDIUM RUSSIAN BATTLE TANKS

The T-34 tanks that turned up unexpectedly in 1941, on the other hand, were desirable booty and were usually adopted at once by the troops capturing them. The armored units used them partly as battle tanks, but they were more often used here too, as in the infantry, by the armored pursuit units, since there were fewer problems of friend/foe recognition when they were used as pursuit tanks. To be sure, the troops still painted large-dimension crosses or swastikas on the tanks, but often enough this could not prevent the tank from being recognized as an enemy by its silhouette and being shot down. Attempts were made to equip captured T-34's with wooden bodies to simulate the silhouettes of German tanks.

When the first T-34 tanks appeared in 1941 and the Germans had nothing similar to use against them, consideration was given to the idea of simply copying the T-34. This was not done, but the construction of the Panther was strongly influenced by the T-34.

Left page:
The medium battle tank with the original designation T-34 (A) was renumbered T-34/76 after the appearance of the T-34/85. It is seen here as type A (in western nomenclature) or as "1941 model" (as it was called in Russia). (BA)

Right:
Tank "A" close-up. Here the antenna is folded backward. The mounting of the bow machine gun, the lights and the towing hook of this 1941 model are easy to see. (BA)

In 1941 the 10th Armored Division captured several T-34 tanks. So that nobody would make off with the usable tanks, they were immediately marked with chalk. An aviation symbol has been tied to the welded turret of this T-34 to indicate its ownership. (BA)

Upper left:
A swastika flag has been tied around the turret hatch here too. The cast turret (photo) had 52-mm armor, the welded version only 45 mm. (BA)

Above:
This T-34 (1941 model) with cast turret is in service with the 10th Armored Division. The tank has not yet been equipped with a radio antenna. (BA)

Left:
Two more T-34/76 A tanks of the 98th Infantry Division. Both tanks have cast turrets. The painted crosses on the hulls differ slightly. (BA)

Painted winter white is this T-34/76 A of the 98th Infantry Division. The tank has accurately painted German crosses on the bow, sides and turret hatch. A letter "A" is also painted on the right track cover. (BA)

A T-34/76 with winter paint in the winter of 1943/44. The German cross is painted inside a circle. (BA)

Above:
Another T-34 of the 98th Infantry Division. The tank is a 1941 type with welded turret, on which a German cross has been painted. (BA)

Upper right:
A tank of the 10th armored division. The number 5 on the turret may be a sequential marking in the "T-34 Column" of Armored Regiment 7. The swastika flag is attached to the rear. Troop engineers are looking for a path without mines. (BA)

Right:
The 9th Armored Division used this T-34 (1941/42 model). Besides large German crosses on the bow and turret, a swastika has been painted on the hatch opening as a friend/foe marking for the Luftwaffe.

Upper left:
A T-34/76 of the 1942 production series in the Russian steppes. The 1942 model—also called T-34/76—can be recognized from behind by the round service hatch, which is rectangular on earlier models. The tank with the striking German cross appears to be stripped, as almost all built-on parts and even the exhaust pipes are missing.

Above:
Here is a T-34/747(r) armored battle vehicle, according to the official German designation. In front of it is an agricultural tractor, which was often used by the troops for towing.

Left:
A T-34/76 with a readily visible German cross on the lid of the turret hatch. Crosses were also presumably painted on the sides of the turret.

Above:
The 3rd SS Division "Totenkopf" also had T-34 tanks in service. Here are two T-34/76 (1941/42 model) tanks under a canvas cover. Along with the German cross on the bow, the division's emblem (the death's-head) can be seen on the turret.

Upper right:
This T-34 (1941/42 model) is completely equipped with all-steel road wheels. The winter-white vehicle is marked with crosses on the hull and the turret sides. (Frank)

Right:
This T-34 was in service with the 18th Armored Grenadier Division in March of 1943. Cylindrical containers for spare fuel can be seen on the tail. The crew has hung their steel helmets on the turret.

Upper left:
A T-34/76 with winter camouflage. Here they are probably testing what size trees can be knocked down. (Frank)

Above:
This T-34/76 A was used by the 1st Armored Division. The large German cross is painted to the rear of the turret sides.

Left:
This T-34/76 also has crosses painted in circles. It can be seen here that it also has a similar cross on the back of the turret. (BA)

T-34/76 1943 Model

(Types D-F)

Left and above:

During the course of 1942, the 1943 model of the T-34 went into production. As Type D, it had a larger hexagonal turret with two turret hatches. The photos show details of the tail, engine cover and turret well. Here a towing procedure is shown.

After one tank is not sufficient, a second T-34 is applied to tow the valuable booty in. The two towing tanks are also newly captured and as yet bear no German markings. While the front and left tanks have the new type of bow machine-gun mount, the right one still has the old type (2 x BA)

Another variation among captured T-34 tanks of the 2nd SS Division is shown by this tank, which was fitted with aprons like those of the German battle tanks. Along with the cross, the turret number 1021 has been painted on, as well as a swastika on the turret hatch for recognition. This photo served as a model for the color drawing on the back cover.

Upper left:
This T-34 took a direct hit at the right, next to the driver's hatch, and an armor plate was welded on there. The tank is equipped with a German Notek camouflage light. (BA)

Above:
The T-34 (1943 model) was also called T-43 by the German troops. This one is said to have been captured by the 18th Armored Grenadier Division. The order of the road wheels is typical. The 1st and 5th road wheels are equipped with rubber tires, since strong vibrations occurred in vehicles with five all-steel wheels.

Left:
The Division "Das Reich" also used captured T-34 tanks in Operation "Citadel". This picture was taken during it.

Liberally covered with recognition markings is this T-34. It has German crosses
on the bow, the sides of the hull and turret, and—as the opened hatches show—a
gigantic cross on the turret roof too. (BA)

Sign reads: Confiscated by the SS Armored Corps. SS Tank Works. Information: J Squadron, plus 2 emblems.

In the spring of 1943 the SS Armored Corps confiscated a large building in recaptured Kharkov. There the Armored Pursuit Unit "Das Reich" repaired captured T-34 tanks and organized them into their own companies.

Upper right:
Captured T-34 tanks were completely dismantled there, overhauled and assembled again as if on an assembly line. So that the parts fitted together correctly, the tanks were numbered and the individual parts had these numbers applied. In back are four T-34 (1943 model) tanks, in the center a T-43 turret and in front a T-34 (1941 model).

Right: A finished T-34, painted with a German cross, leaves the "factory". At the front of the picture is a KW-I. (3 x BA)

T-34 (1943 model) with its new crew. This tank also belonged to the SS Division "Das Reich". Since these tanks were prepared in the "factory", they all had the same built-on parts and German Notek camouflage lights. (Munin)

Above:
This T-34's markings can be seen despite its camouflage. Along with the crosses and numbers on the turret sides, the light-colored insides of the turret hatches are also painted with crosses. At the 6-o'clock turret position they serve, as here, as recognition signs from forward, and in an attack one can recognize one's own tank from the rear. (Munin)

Upper right:
Again, four T-34 tanks of the Division "Das Reich", including 1943 models, easy to recognize by the turrets with the two round hatches. (Munin)

Right:
The SS Division "Das Reich" used the tanks in mixed companies with Tank IV units. Here, along with the German tanks, are four T-34's, including one 1941 and three 1943 models. (Munin)

The Armored Pursuit Detachment 38 of the 2nd Armored Division used these two T-34 (Type D) tanks as "pursuit tanks". There were not so many problems with friend-for recognition when captured tanks were used as pursuit tanks.

To give the commander a better panoramic view, the cupola of a German Panzer III or IV tank was built onto this T-34. In the middle of the bow armor, the number 20 can be seen. (BA)

This T-34/76 (Type E) served as a model for our cover picture.

Upper left:
The later versions of the T-34 (1943 model) were equipped by the Russians with a simple commander's cupola. As with the earlier versions, there were either cast round or welded hexagonal turrets, depending on the factory in which they were made. (BA)

Above:
This T-34 has been captured intact by German troops and is now being tested. German markings have not yet been painted on. (BA)

Left:
In the winter of 1943/44 Armored Detachment 21 of the 20th Armored Division was able to capture a large number of T-34 (1943 model) tanks. (BA)

This T-34 was put into service by the Armored Grenadier Division "Grossdeutschland". The division emblem (a white steel helmet) can be seen clearly on the turret. (BA)

The T-34 tank with its 76-mm cannon was welcome booty. These were immediately marked as German tanks with air-identification cloths. The front tank has the old-type rubber-covered road wheels, the others have the newer type. (BA)

The leading tank of the 2nd Column, with turret number 221. To the right near the driver's hatch a German cross can be seen. Since these tanks were in service as pursuit tanks, the main identifying mark is on the rear. (Cutshorn)

Above:
This tank was marked very clearly as a German vehicle. Around the turret, big white crosses alternate with swastikas. (BA)

Upper right:
Ammunition is piled up next to Tank 231. Some 80 cartridges could be carried. (Weber)

Right:
In addition to T-34's with cast turrets, there were also tanks with welded turrets in this unit, as here number 211. Some captured tanks were still in service with Armored Pursuit Detachment 128 in 1945. (Weber)

Left page:
Captured and still usable Russian tanks—here a T-34 (E)—were prepared for service in German units by covering holes (note the armor plate in the middle of the black cross), being equipped with radios (antenna) and having sequential numbers painted on their turrets.

Above:
The big German crosses were painted carefully with stencils. The edges of the crosses were drawn on with chalk and then painted by hand.

Right:
The tanks were given the three-color camouflage paint typical of the time (see front cover). The turret numbers were also painted on with stencils. (BA)

T-34/85

Built from the end of 1943 on, the new T-34/85 with a three-man turret and the new 85-mm cannon appeared at the front in 1944. This tank at the Warsaw-Jadow intersection shows a white German cross at right near the driver's hatch and a swastika flag. It was captured and put in service by the 5th SS Armored Division "Wiking" in 1944. (BA)

Left:
These two T-34/85 tanks were captured by the 252nd Infantry Division in East Prussia during 1944. The tanks were made recognizable as German vehicles with crude black crosses and put in service with Armored Pursuit Detachment 252. There were only a few captured T-34/85 tanks, since they appeared on the battlefield only at a time when the German troops were almost always retreating—and so few were captured.

HEAVY RUSSIAN BATTLE TANKS

As with the T-28 before it, nothing is known of any reuse of a T-35, likewise a multiturreted giant tank. The heavy KW-I battle tanks, on the other hand, were as welcome booty as the T-34. They too were immediately taken possession of and put into service by the capturing troops.

When in May of 1942 a company was to be organized with captured Russian heavy tanks for the landing on Malta ("Operation Hercules"), this could be done only under great difficulty, since most of the heavy tanks had been destroyed and the usable ones had already been put in service with the troops.

When in 1944 the heavy "Josef Stalin" tanks appeared at the front, the Wehrmacht's days of capturing tanks were over because of their constant retreats, and there was great difficulty in taking even one example for the Army Weapons Office, which naturally was interested in examining this tank.

KW-I

A KW-I of the first production batch (1939 model), with a 76-mm cannon, Machanov L-11. The vehicle bears only a small German cross and the noticeable number H 03. The tank has been equipped with a German commander's cupola. At the left front of the track cover is the typical German auto jack. This tank was put in service with Armored Regiment 1. (Riebenstahl)

Right:
This KW-I (1940 model) loaded on a railroad car bears huge German crosses on its bow and turret sides. Another eye-catching identifying mark is the colored stripe painted around the turret. The 1940 model was provided with heavier turret armor.

Left page:
This KW-I (1942 model) was in service with the 1st Armored Division. The 1942 model was also equipped with a ZIS-5 cannon, but had only a cast turret. (Riebenstahl)

The three KW-I tanks shown on this page were also in service with Armored Regiment 1 of the 1st Armored Division—here still without German crosses. (Riebenstahl)

Upper left:
Here a T-34 is being towed by a KW-I (1942 model).
The two tanks were captured by the 8th Armored
Division but not yet marked with German crosses.

Above:
This KW-I (1940 model) was also in service with the 8th
Armored Division. It belonged to the 1st Company of
Armored Regiment 10. The 1940 model was armed with
the 76-mm Grabin F-32 cannon. These tanks were used
by the Germans under the designation "Armored Battle
Vehicle IA 753(r)".

Left:
The Armored Regiment 204 of the 22nd Armored
Division captured a KW-I (1941 model), with the longer
ZIS-5 cannon, near Byelogrod at the end of 1943, and
went to much trouble to rebuild the tank. (Brändlin)

From the KW-I shown at the bottom of the lefthand page there arose this KW-I with a German 7.5-cm L/43 cannon and German commander's cupola. The armor was sprayed with German camouflage paint. (Brändlin)

KW-II

Above:
More an artillery carrier than a battle tank was the KW-II with its 152-mm howitzer and 53-ton weight. In the German Army it was designated Armored Battle Vehicle KV-II 754(r). This badly shot-up example was at the Krupp factory in Essen in 1945.

Right:
This KW-II bears a German cross on the left side of its turret. Only a few of these giants with the 12-ton turret were used by the German Army, as they were scarcely to be towed.

Much time and trouble were devoted to these giants; this one was even given a German commander's cupola. The rear door of the turret is open while driving, and there appear to be racks for additional ammunition on the rear. (BA)

AFTERWORD

If this volume should find acceptance among friends of the series, further volumes on captured battle tanks from other countries, but also on pursuit tanks, armored artillery and armored halftracked and wheeled vehicles are possible. There is no lack of material. The editors and author would be happy to make contact in this way with former members of captured-tank troop units.

ALSO FROM SCHIFFER MILITARY

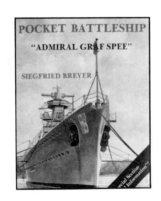